AF555962

Young Learner's

MENTAL MATHS

Book-5

Rajesh Singh

CONTENTS

Published by:

YOUNG LEARNER PUBLICATIONS®

G-1A Rattan Jyoti, 18 Rajendra Place
New Delhi-110008 (INDIA)
Tel.: 011-25750801, 25820556, 25755559
Fax: 91-11-25764396
Email: goodwillpub@gmail.com
gph.ylp@gmail.com
gph.ylp@goodwillpublishinghouse.com
Website: www.goodwillpublishinghouse.com

Worksheet - 1

Numbers, Numerals, Place Value Charts

1. **Write in millions, hundred thousands, ten thousands, thousands, hundreds, tens and ones.**

   2 345 907 ____

   ____

2. **Write in ten millions, millions, hundred thousands, ten thousands, thousands, hundreds, tens and ones.**

   98 734 215 ____

   ____

3. **Write the numeral for each of the following:**

   a) 3 millions 7 hundred thousands 6 ten thousands 5 hundreds 9 tens and 1 one. ____________________

   b) 2 ten millions 6 millions 4 hundred thousands 2 ten thousands 7 thousands 3 tens and 6 ones. ____________________

4. **Write the following numbers in place value chart:**

		TM	M	H-Th	T-Th	Th	H	T	O
a)	8 373 958								
b)	89 347 927								

5. **Write the numbers represented in following place value chart in figures:**

	TM	M	H-Th	T-Th	Th	H	T	O
a)		3	2	4	1	3	6	8
b)	8	2	0	2	2	3	8	8

a) ____

b) ____

Worksheet - 2

Number Name, Expanded Form, Successor and Predecessor

1. Write the number name for each of the following:

a) 5 239 801 ________________________________

____

b) 87 297 258 ________________________________

____

2. Write the numeral for each of the following using commas:

a) Seven million seven hundred thirty-eight thousand four hundred one

____

b) Thirty million one hundred twenty thousand two hundred eighty

____

3. Write the following in expanded form:

a) 7214750 = ________________________________

____

b) 29909909 = ________________________________

____

4. Write the numeral for each of the following:

a) 2000000 + 500000 + 80000 + 200 + 70 + 6 = ________________

b) 90000000 + 900000 + 40000 + 7000 + 10 + 9 = ________________

5. Write the successor of 1287451 ________________

6. Write the next two numerals for 12764399 ________________

7. Write the predecessor of 75212093 ________________

8. Write the two numerals that come just before 7012400 ________________

9. Write two numbers that come in between:

5647248 ________________________________ 5647251

10. a) Counting by two millions write two numerals from 1 200 000.

____

b) Counting by ten millions write two numerals from 23 000 000.

____

Worksheet - 3

Greater and Smaller Number
Ascending and Descending Order, Rounding-off Numbers

1. Underline the greater number: 6446316 and 6442216
2. Underline the smaller number: 19443741 and 19453748
3. Arrange the following in ascending order:

   5484572, 5477997, 5546499 ____________________
4. Arrange the following in descending order:

   24273627, 24285727, 24277420 ____________________
5. Round 6775414 to nearest million ____________________
6. Round 27654258 to nearest ten million ____________________
7. a) Write the largest and the smallest numerals using each of the digits 1, 2, 3, 4, 5, 6, and 7. Repetition is not allowed.

   Largest: ____________ Smallest: ____________

   b) Write the largest and the smallest 8-digit numerals using the digits 0, 1, 2, 3, 4, 5, 6, 7 and 8. Repetition is allowed.

   Largest: ____________ Smallest: ____________
8. Largest 7 digit number using all digits except 3, 5 and 7 ____________
9. Smallest 8 digit number with 2's and 5's appearing alternately ____________
10. Largest number less than 3578000 ____________________
11. Smallest number more than 28751999 is ____________________.
12. Largest number less than 4157580 and having all digits different is ________.
13. Smallest 7 digit number with all digits except one as 0's is ____________.
14. Greatest number between 3147542 and 5556669 having all digits same is ____________.

Worksheet - 4

Roman Numerals

1. Fill in the blanks.

a) A number which cannot be represented in Roman system is ___________.

b) The basic symbols in the Roman system are ___________.

c) The Roman numerals that cannot be subtracted are ___________.

d) The Roman numerals that can be subtracted from C are ___________.

e) MMM…. (10 times) represents ___________.

2. Write the following in Roman numerals:

a) 48 ___________ b) 97 ___________

c) 182 ___________ d) 363 ___________

e) 624 ___________ f) 735 ___________

g) 819 ___________ h) 1356 ___________

3. Tick the ones which are meaningful and cross-out the ones which are meaningless.

DDXV CMLIX LVXIII CDLXVI VXII MMDXL

4. Write the following in Indo-Arabic numerals:

a) XXV ___________ b) XLVI ___________

c) CXXXIX ___________ d) CCXVIII ___________

e) DCLXI ___________ f) MDCLXIII ___________

g) MDCXCIII ___________ h) MMDCCXIV ___________

5. Fill in the blanks.

a) LXV + ___________ = CXXX b) ___________ + CDLX = M

Worksheet - 5

Adding and Subtracting

1. Add the following:

a)

	2	8	2	2	0	1	6
+	3	1	3	4	9	7	1

b)

	2	5	1	2	5	5	2	4
+	6	1	5	2	1	2	3	5

c)

	4	8	3	4	6	9	2
+	1	2	5	7	8	9	0

d)

	1	4	3	4	6	7	4	7
+	5	2	9	5	6	8	9	8

2. Subtract the following:

a)

	9	4	7	5	3	8	4
–	4	1	3	4	2	7	3

b)

	6	6	7	4	8	9	7	4
–	1	5	6	2	5	5	7	1

c)

	9	4	8	3	5	2	6
–	7	8	8	4	7	3	7

d)

	7	0	6	2	4	2	6	3
–	6	6	8	9	9	6	7	3

Worksheet - 6

Word Problems

1. Population of a city was 10,287,451 two years ago. It has increased by 124,897 since then. What is the population of this city now?

2. In an election Mr. Vincent got 1,271,485 votes, and Ms. Simone got 1,238,441 votes. Also, 190,802 voters did not turn up. How many voters were there in total? What was the victory margin?

3. 1,963,284 tourists visited Paris this year. 1,884,214 tourists had visited Paris last year. How many more tourists visited Paris this year?

4. Mr. Sam had $ 12,497,500 in his bank account. He received a payment of $ 2,085,400. He also paid $ 2,475,500 for the goods he purchased. What is his bank balance now?

Adding and Subtracting Together
Properties of Addition and Subtraction

1. Simplify: 6524743 – 2574184 + 3178231

____

____

____

2. Check the following subtraction by addition:

5 6 2 8 7 4 1	....	Subtraction is
– 2 3 0 5 4 2 0	+	....
3 3 2 3 3 2 1	____________	(correct / incorrect)

3. Check the following addition by subtraction:

6 3 2 5 4 1 7	7 4 4 7 9 8 5	Addition is
+ 1 1 2 2 4 7 8	– 6 3 2 5 4 1 7	....
7 4 4 7 9 8 5	____________	(correct / incorrect)

4. Fill in the blanks.

a) 1223272 + 4323783 = 4323783 + ____________

b) 1233721 + ____________ = 1282372 + 1233721

c) ____________ + 54742632 = ____________ + 47847593

d) (____________ + 5170424) + 330674 = 517004 + (5170424 + 330674)

e) (2146219 + 178463) + 189047 = 2146219 + (____________ + 189047)

f) (2447175 + 4218621) + ____________ = (2447175 + 4218621) + 2531287

g) 45281352 + 0 = ____________

h) ____________ + 0 = 53618679

i) ____________ + 1728435 = 1728435

j) 458352 – 0 = ____________

k) ____________ – 0 = 32351023

l) 1921376 – ____________ = 0

Worksheet - 8

Multiplication

Solve the following:

a)

1	3	4	2	1	0	4
					×	2

b)

1	0	2	3	2	3	3	1
						×	3

c)

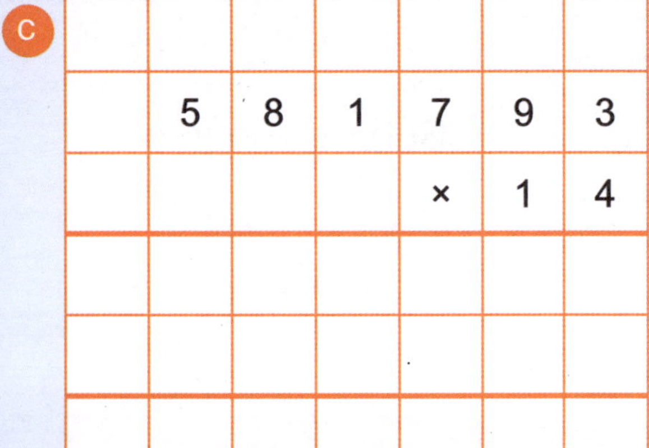

	5	8	1	7	9	3
				×	1	4

d)

		6	4	8	9	4	7
					×	3	7

e)

		5	0	0	0	9
				×	5	9

f)

	1	3	3	3	3	3	5
					×	6	5

g)

		9	9	6	2	1
				×	8	6

h)

		1	4	1	4	0	2
					×	9	9

Worksheet - 9

Long Division With and Without Remainder

Solve the following:

1 4215072 ÷ 6 Q = ___________	2 6523488 ÷ 18 Q = ___________
3 2390890 ÷ 9 Q = ________ R = ________	4 12928345 ÷ 19 Q = ________ R = ________

Worksheet - 10

Word Problems

1. Each packet has 36,450 brushes. How many brushes will 18 such packets have?

____

____

____

2. A lorry can carry 9,275 bags of cement in one round. How many bags can it transport in 28 rounds?

____

____

____

3 8 items can be packed in a box. How many boxes are needed to pack 677,464 items?

________ boxes.

4 4,215,320 items have to be packed in boxes, each containing 50 items. How many boxes can be made? How many items will be left behind?

________ boxes. ________ items.

Worksheet - 11

Multiplying by Multiples of 10
Properties of Multiplication and Division

1. Fill in the blanks.

a) 24132 × 1000 = ______________ b) 653 × 10000 = ______________

c) 3 × 10000000 = ______________ d) 5 × 10000000 = ______________

e) 4000 × 2211 = ______________ f) 211 × 50000 = ______________

g) 330 × 4000 = ______________ h) 45000 × 30 = ______________

i) 7000 × 1600 = ______________ j) 3200 × 3300 = ______________

k) A truck can carry 500,000 pouches of salt in one round. How many pouches will it carry in 12 rounds?

____

l) Each carton has 20,000 toffees. How many toffees will 600 such cartons have?

____

2. Solve the following:

a) 9642458 × 1 = ______________

b) ______________ × 1 = 7468521

c) 342564 × ______________ = 3485647 × 342564

d) 874141 × 96424 = 96424 × ______________

e) 456120 × ______________ = 123908 × ______________

f) 321457 × [621248 × 32341] = [321457 × ______________] × 32341

g) 5325 × 58743 × 9541232 = [___________ × ___________] × 9541232

h) 123400 × [1234000 × __________] = [__________ × __________] × 1234

i) 9621472 × ______________ = 0 j) 800002 × 2309 × __________ = 0

k) 2906000 ÷ 1 = ______________ l) 0 ÷ 12000000 = ______________

Worksheet - 12

Estimating Sum, Difference, Product and Quotient
Estimation Stories

1. **Estimate by rounding to nearest millions.**

   a) 2521722 + 5243783 ≈ __________ + __________ = __________

   b) 6241541 – 3857421 ≈ __________ + __________ = __________

2. **Estimate by rounding to nearest ten millions.**

   a) 13104286 + 25241854 ≈ __________ + __________ = __________

   b) 23251471 – 12214512 ≈ __________ + __________ = __________

3. **Estimate the product by rounding to nearest hundreds.**

   9842 × 978 ≈ __________ × __________ = __________

4. **Estimate the product by rounding to nearest thousands.**

   14175 × 998 ≈ __________ × __________ = __________

5. **Estimate the following by rounding to nearest tens.**

   715 ÷ 93 ≈ __________ ÷ __________ = __________

6. **Estimate the following by rounding to nearest hundreds.**

   126419 ÷ 780 ≈ __________ ÷ __________ = __________

7. **Estimate the following by rounding to nearest thousands.**

   880234 ÷ 8168 ≈ __________ ÷ __________ = __________

8. **Sonia has a bank balance of $ 2,142,351. Her brother has a bank balance of $ 3,362,514. Find the difference in their bank balances rounded to nearest millions.**

   ____________________ – ____________________ = ____________________

9. **Population of a country is 23,456,908. Population of a neighbouring country is 38,956,700. Find the total population of the two countries rounded to nearest ten millions.**

   ____________________ + ____________________ = ____________________

Worksheet - 13

DMAS and BODMAS

Simplify the following:

1. 13 – 2 × 6 ÷ 2 + 1

2. 30 ÷ 5 + 2 × 6 – 3

3. 10 + 4 × 2 – 5 ÷ 5

4. 4 × [6 – {15 ÷ (3 × 2 – 1) + 2} + 2]

5. 24 – [4 × {15 – (6 ÷ 2 × 3)} – 3]

Worksheet - 14

Multiples of Numbers
Even and Odd Numbers

1. **Write the first five multiples of 14** ____________________

2. **Write the multiples of 9 which lie between 30 and 70** ____________

3. **Tick yes or no for each of the following:**

   a) 145 is a multiple of 15

   ☐ Yes ☐ No

   b) 144 is a multiple of 18

   ☐ Yes ☐ No

4. **Cross out the incorrect multiple(s) in the following:**

   19, 38, 59, 76, 95, 113, 134, 152, 171, 190

5. **Fill in the blanks.**

   a) An even number ends in ____________________.

   b) The smallest even number is ____________________.

   c) An odd number ends in ____________________.

   d) A number which is not a multiple of __________ is called an odd number.

6. **Tick (✓) the odd numbers, cross-out (×) the even numbers.**

   9013, 1265, 9826, 12188, 23110

7. **Write the first five two-digit even numbers** ____________________

8. **Write first five three-digit odd numbers** ____________________

9. **Write all odd numbers between 3096 and 3107** ____________________

10. **Give first three even multiples of 11** ____________________

11. **Write true or false.**

    a) Multiples of an odd number need not be odd ____________________

    b) Multiples of an even number need not be even ____________________

Worksheet - 15

Factors and their Properties
Divisible Numbers

1. Fill in the blanks.

a) When we multiply any two or more numbers, we get a product. Each number is called a _____________ of the product.

b) If 8 × 12 = 96, then 8 and 12 are factors of ______________.

c) If a × b × c = d, then _____, _____ and _____ are factors of _____.

d) If 10 × 8 × 3 = 240, then _____, _____ and _____ are factors of 240.

e) Every number is a ___________ of itself.

f) Factors of 9 are _______________________.

g) Factors of 45 are _______________________.

h) Odd factors of 40 are _______________________.

i) If 72 is divisible by 9, then 9 is a _______________ of 72.

j) If 63 is divisible by ______, then 7 is a factor of 63.

k) If 104 is divisible by _____ and _____ both, then 13 and 8 are factors of _____.

l) If 28 is divisible by 7, then ______ is a factor of ______.

2. Write true or false.

a) If 17 × 11 = 187, then 11 and 17 are factors of 187. ______

b) If 11 × 3 × 10 = 330, then 11, 3 and 10 may not be the factors of 330. ______

c) If 7 × 5 = 35 and 35 × 11 = 385, then 7, 5 and 11 are factors of 385. ______

d) If 2 × 4 × 6 × 8 = 384, then 24 is not a factor of 384. ______

e) An even number will always have 2 as a factor. ______

f) An odd number can have an even factor. ______

g) All multiples of 11 will have 11 as a factor. ______

h) All numbers have two or more factors. ______

Worksheet - 16

Prime and Composite Numbers
Rules of Divisibility

1. Fill in the blanks.

a) A number that has exactly two factors is called a __________ number.

b) There are __________ prime numbers less than 50.

c) There are __________ prime numbers less than 100.

d) '2' is the only number which is both ____________ and ____________.

e) A pair of consecutive prime numbers is ____________.

f) Express 30 as the difference of two prime numbers: 30 = ______ – _______

g) Find the greatest prime number which is less than 44 ____________

h) A number which has more than two factors is called a __________ number.

i) Composite numbers between 90 and 100 are ______________________.

2. Encircle the numbers that are;

a) divisible by 2: 4451, 1962, 2413, 9464, 8525, 3367

b) divisible by 3: 5141, 2029, 3153, 4213, 8296, 9327

c) divisible by 4: 1002, 8498, 7544, 3116, 5594, 7636

d) divisible by 5: 4435, 5380, 7841, 26502, 83336, 51445

e) divisible by 6: 2434, 4392, 3192, 4231, 16270, 33432

f) divisible by 9: 4230, 5292, 7065, 4140, 9092, 9453

g) divisible by 10: 7410, 9233, 4520, 4446, 5460, 6374

3. The value(s) of 'a' for which 635a is divisible by 2 is (are) ____________.

4. The value(s) of 'a' for which 4a12 is divisible by 3 is (are) ____________.

5. The value(s) of 'a' for which 130a is divisible by 4 is (are) ____________.

6. The value(s) of 'a' for which 330a is divisible by 5 is (are) ____________.

7. The value(s) of 'a' for which 6a20 is divisible by 10 is (are) ____________.

Worksheet - 17

Prime Factorization

Express each of the following numbers as a product of prime factors using a factor tree:

1 8	2 28
3 52	4 72
5 80	6 100

Worksheet - 18

HCF and LCM

1. **Find the HCF of the following:**

a 36 and 48	b 30, 60 and 90

2. **There are 120 plain cookies and 140 almond cookies. They are to be packed with same number and same type of cookies in each packet. What is the highest number of cookies that can be packed in one packet?**

____

____

3. **Find the LCM of the following:**

a 15 and 18	b 20, 30 and 40

4. **Two brands of chocolates are available in packs of 24 and 15 respectively. If I need to buy an equal number of chocolates of both kinds, what is the least number of boxes of each kind I would need to buy?**

____

____

5. **Three bells toll at intervals of 15 minutes, 10 minutes and 20 minutes. If they toll together at 11.00 am, when do they toll together next?**

____

____

Unit-4 Fractions

Worksheet - 19

Types of Fractions

1. Tick the improper, and cross-out the proper fractions.

$\frac{9}{19}$ $\frac{11}{2}$ $\frac{25}{7}$ $\frac{3}{9}$ $\frac{8}{17}$ $\frac{105}{46}$ $\frac{199}{220}$

2. Write the following as mixed fractions:

a) Two and one-fourth: ______ b) 3 and $\frac{8}{9}$: ______ c) $55 + \frac{1}{56}$: ______

3. Express the following as improper fractions:

a) $3\frac{5}{6} =$ b) $2\frac{3}{5} =$ c) $4\frac{3}{8} =$

d) $2\frac{5}{7} =$ e) $11\frac{1}{7} =$ f) $3\frac{1}{9} =$

4. Express the following as mixed fractions:

a) $\frac{17}{6} =$ b) $\frac{11}{8} =$ c) $\frac{13}{9} =$

d) $\frac{22}{7} =$ e) $\frac{100}{9} =$ f) $\frac{200}{13} =$

5. Which of the following are pairs of like fractions? Encircle them.

a) $\frac{13}{9}$ and $\frac{5}{9}$ b) $\frac{3}{5}$ and $\frac{12}{5}$ c) $\frac{2}{13}$ and $\frac{4}{3}$

d) $\frac{14}{141}$ and $\frac{41}{141}$ e) $\frac{1}{6}$ and $\frac{11}{66}$ f) $\frac{79}{78}$ and $\frac{78}{79}$

6. Reduce the following to lowest terms:

a) $\frac{135}{75} =$ b) $\frac{18}{33} =$ c) $\frac{54}{90} =$

d) $\frac{196}{14} =$ e) $\frac{1010}{1110} =$ f) $\frac{29}{89} =$

Worksheet - 20

Addition, Subtraction, Multiplication and Division of Fractions

1. Add the following:

a) $\frac{2}{4} + \frac{1}{4}$ = ________ = ________

b) $\frac{5}{10} + \frac{2}{10} + \frac{1}{10}$ = ________ = ________

c) $\frac{1}{2} + \frac{2}{5}$ = ________ = ________

d) $\frac{4}{7} + \frac{3}{14} + \frac{3}{10}$ = ________ = ________

2. Subtract the following:

a) $\frac{3}{5} - \frac{1}{5}$ = ________ = ________

b) $\frac{9}{10} - \frac{4}{10} - \frac{3}{10}$ = ________ = ________

c) $\frac{1}{3} - \frac{1}{24}$ = ________ = ________

d) $\frac{5}{8} - \frac{1}{6} - \frac{1}{4}$ = ________ = ________

3. Solve the following:

a) $\frac{4}{14} + \frac{1}{2} - \frac{5}{7}$ = ________ = ________

b) $\frac{2}{22} + \frac{9}{11} - \frac{1}{2}$ = ________ = ________

c) $\frac{16}{17} - \frac{1}{289} + \frac{3}{17}$ = ________ = ________

d) $\frac{44}{84} - \frac{21}{42} + \frac{3}{21} - \frac{1}{6}$ = ________ = ____

4. Multiply the following:

a) $\frac{5}{8} \times 2$ = ________

b) $9 \times \frac{3}{8}$ = ________

c) $\frac{3}{7} \times \frac{2}{3}$ = ________

d) $\frac{5}{17} \times \frac{7}{15}$ = ________

5. Divide the following:

a) $5 \div \frac{2}{5}$ = ________ = ________

b) $\frac{3}{8} \div 2$ = ________ = ________

c) $\frac{3}{13} \div \frac{7}{15}$ = ________ = ________

d) $\frac{4}{27} \div \frac{13}{9}$ = ________ = ________

Worksheet - 21

Word Problems

1. Rosy did $\frac{1}{4}$ of her homework on Monday and $\frac{3}{8}$ of it on Tuesday. What part of homework was done over the two days?

2. From a chord $\frac{8}{9}$ m long, a piece $\frac{4}{5}$ m was cut off. What is the length of the chord left?

3. $\frac{2}{5}$ of the school garden is looked after by students of Class V. Of this part, $\frac{3}{4}$ is used to raise flowers. What fraction of the school garden is used to raise flowers?

4. 8 big sheets were cut into smaller pieces such that each piece is $\frac{1}{8}$ of the original sheet. How many small pieces were obtained?

Unit-5
Decimals

Worksheet - 22

Tenths, Hundredths and Thousandths, Place Value Chart
Expanded Form

1. How will you read the following?

a) 0.5 ____________________

b) 10.06 ____________________

c) 2.043 ____________________

2. Write the following fractions as decimals:

a) $\frac{2}{10}$ = ____________ b) $1\frac{3}{10}$ = ____________

c) $\frac{21}{100}$ = ____________ d) $1\frac{9}{100}$ = ____________

e) $\frac{7}{1000}$ = ____________ f) $3\frac{123}{1000}$ = ____________

3. Write the following decimals as fractions or mixed numerals:

a) 0.2 = ______ b) 1.43 = ______ c) 1.463 = ______ d) 2.014 = ______

4. Represent the following decimals on the place value chart:

a) 6.2 b) 18.57 c) 123.052

	Hundreds 100	Tens 10	Ones 1	Tenth $\frac{1}{10}$	Hundredths $\frac{1}{100}$	Thousands $\frac{1}{1000}$
a)						
b)						
c)						

5. Write the decimal for the following:

a) $1 + \frac{1}{10}$ = ________ = ________ b) $1 + \frac{4}{10} + \frac{6}{100}$ = ________ = ________

c) $7 + \frac{1}{10} + \frac{7}{100} + \frac{6}{1000}$ = ____________ = ____________ = ____________

6. Expand the following:

a) 4.6 = ____________________ b) 1.02 = ____________________

Worksheet - 23

Like and Unlike Decimals
Adding and Subtracting Decimals

1. Tick the correct type of decimal.

a) 4.36, 0.74 Like decimals ☐ Unlike decimals ☐

b) 5.367, 17.33 Like decimals ☐ Unlike decimals ☐

2. Write True or False.

a) 0.3 = 0.30 ____________ b) 0.2 = 0.02 ____________

c) 0.41 = 0.041 ____________ d) 4.09 = 4.900 ____________

3. Express each of the following pairs as like decimals:

a) 4.3, 0.84 ____________ b) 23.62, 2.937 ____________

c) 1.9, 0.002 ____________ d) 5.87, 7.9 ____________

4. Add the following:

a) 2.9 and 12.7	b) 2.98 and 5.873

5. Subtract the following:

a) 2.9 from 10.7	b) 2.098 from 10.907

Worksheet - 24

Multiplying and Dividing Decimals
Dividing a Decimal by Multiples of 10

1. Multiply the following:

a 5.24 by 6	b 0.123 by 17

2. Divide the following:

a 55.8 by 6	b 77.16 by 12
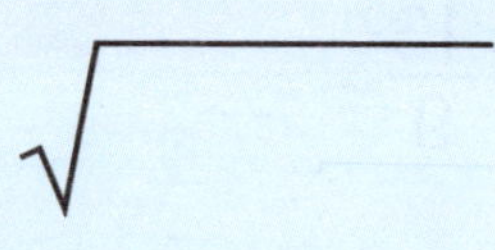	

3. Solve the following:

a) $7.52 \times 10 =$ ____________ b) $4.916 \times 100 =$ ____________

c) $3.934 \times 1000 =$ ____________ d) $6.2 \div 10 =$ ____________

e) $224.8 \div 100 =$ ____________ f) $12.5 \div 1000 =$ ____________

Worksheet - 25

Word Problems

Fill in the blanks.

1. The length of a line segment is 6.3 cm. What is its length in millimetres? __________ mm
2. A piece of ribbon is 2.95 m long. What is its length in centimetres? ________ cm
3. A wire is 3.08 m long. What is its length in cm? __________ cm
4. Two stops on a metro route are 2.315 km apart. What is the distance between them in metres? __________ m
5. A canal is 12.934 km long. What is its length in metres? __________ m
6. A road is 2.55 km long. What is its length in metres? __________ m
7. A path is 1.08 m long. What is its length in cm? __________ cm
8. A baby weighs 3.995 kg. What is the weight in grams? __________ g
9. A courier packet weighs 4.65 kg. What is the weight in grams? __________ g
10. A pouch weighs 0.296 kg. What is its weight in grams? __________ g
11. A bottle has a capacity of 3.052 L. How many millilitres of water can it hold? __________ ml
12. A bucket can hold 12.8 L of water. What is its capacity in millilitres? ________ ml
13. A pipe is 8 m 19 cm long. Express its length in metres. __________ m
14. Tom bought 3 kg 490 g of tea. Express the weight of tea in kilograms. _____ kg
15. A tank can hold 41 L 650 ml of water. Express its capacity in litres. ________ L

Worksheet - 26

Standard Units of Measurement

Fill in the blanks.

1. The standard unit of length is the ___________. Its symbol is ___________.
2. A metre is divided into 100 equal parts. Each part is called a ___________. Its symbol is ___________.
3. ___________ centimetres make 1 metre.
4. A centimetre is divided into 10 equal parts. Each part is called a ___________. Its symbol is ___________.
5. ___________ millimetres make 1 centimetre.
6. A kilometre is divided into 1000 equal parts. Each part is called a ___________. Its symbol is ___________.
7. ___________ metres make 1 kilometre.
8. The standard unit of weight is the ___________. Its symbol is ___________.
9. A kilogram is divided into 1000 equal parts. Each part is called a ___________. Its symbol is ___________.
10. ___________ grams make 1 kilogram.
11. The standard unit of capacity is the ___________. Its symbol is ___________.
12. A litre is divided into 1000 equal parts. Each part is called a ___________. Its symbol is ___________.
13. ___________ millilitres make 1 litre.
14. To convert smaller unit to bigger unit we ___________ (multiply/divide).
15. To convert bigger unit to smaller unit we ___________ (multiply/divide).
16. ___________ cm make one decimetre.
17. 100 kg make one ___________.
18. ___________ kg make one ton.
19. ___________ litres make one kilolitre.

Worksheet - 27

Converting Units of Length, Weight and Capacity

Fill in the blanks.

1. 40 mm = ______________ cm
2. 60 cm = ______________ dm
3. 120 dm = ______________ m
4. 46000 m = ______________ km
5. 59 cm = ______________ mm
6. A rope is 8 m long. What is its length in dm? ______________ dm
7. Distance between two bus stops is 5000 m. What is the distance in km? ______________ km
8. 4000 mg = ______________ g
9. 5000 g = ______________ kg
10. 900 kg = ______________ q
11. 600 q = ______________ ton
12. 24 g = ______________ mg
13. A cupboard weighs 3 quintals. What is its weight in kg? ______________ kg
14. A truck weighs 6 tons. What is its weight in kg? ______________ kg
15. 4000 ml = ______________ L
16. 2 L = ______________ ml
17. 5000 L = ______________ kl
18. Capacity of a jug is 2 L. What is its capacity in ml? ______________ ml
19. Capacity of a water tank is 2000 L. What is its capacity in kl? ______________ kl
20. A bucket can hold 18000 ml of water. What is its capacity in L? ______________ L
21. Capacity of a water tank is 6 kl. What is its capacity in L? ______________ L
22. A water tank can hold 23000000 ml of water. What is its capacity in kl? ______________ kl

Worksheet - 28

Addition of Metric Measures

Add the following:

1 27 cm and 48 cm	2 4.28 m and 3.09 m	3 12.098 km and 1.94 km
4 560 ml and 235 ml	5 2.98 L and 3.86 L	6 12.98 kl and 13.019 kl
7 237 mg and 632 mg	8 23.674 g and 7.981 g	9 12.982 kg and 14.98 kg

Worksheet - 29

Subtraction of Metric Measures

Subtract the following:

1 67 cm from 91 cm	2 1.67 m from 2.9 m	3 43.98 km from 100 km
4 235 ml from 800 ml	5 4.9 L from 6.783 L	6 2.98 kl from 3.109 kl
7 108 mg from 400 mg	8 12.9 g from 20 g	9 15.9 kg from 20.005 kg

Worksheet - 30

Multiplication of Metric Measures

Multiply the following:

1 75 cm by 5	2 2.08 m by 8	3 23.6 km by 10
4 275 ml by 11	5 35.14 L by 14	6 3465 kl by 15
7 126 mg by 18	8 780 g by 20	9 7125 kg by 25

Worksheet - 31

Division of Metric Measures

Divide the following:

1 126 cm by 6	2 23.4 m by 8	3 235 km by 10
4 996 ml by 12	5 87.5 L by 14	6 204.8 kl by 16
7 187 mg by 17	8 2350 g by 20	9 7896 kg by 24

Worksheet - 32

Word Problems – I

1. A highway is 234 km long and another highway is 697 km long. What is the length of the two highways taken together?

2. A road 96.6 km long needs to be built. Out of this, 55.1 km has already been built. What length of the road is yet to be built?

3. Using his pencil as a standard, John found that his room is 40 pencils long. If the length of the pencil is 20 cm, how many metres long is the room?

4. Using her belt as a standard, Lisa found that their living room is 12 belts long. If the living room is 9 m long, then how many centimetres long is her belt?

Worksheet - 33

Word Problems – II

1. Three shipment containers weigh 6000 kg, 20 quintals and 4 tons. What is the total weight of this consignment in tons?

2. A drum had 2 tons of cement. Out of this 1200 kg was consumed. What weight in quintals of cement is left in the drum now?

3. 200 pouches of coffee weigh 50 kg. What is the weight of each pouch in g?

4. 35 grams of medicine was used in making 70 capsules. How many mg of medicine was put in each capsule?

Worksheet - 34

Word Problems – III

1. The capacity of two overhead tanks is 4700 L and 6300 L. What is the total capacity of the two tanks in kl?

2. A container had 6300 L of oil. Due to a leak at the bottom 2300 L was lost. How many kl of oil is left in the container now?

3. Each vial is to have 250 ml of medicine. How many litres of medicine is needed to fill 300 such vials?

4. 28000 ml of hair oil was equally distributed in 14 containers. How many litres of hair oil does each container have?

Worksheet - 35

Time in Hours and Minutes

1. Write the time shown on each clock.

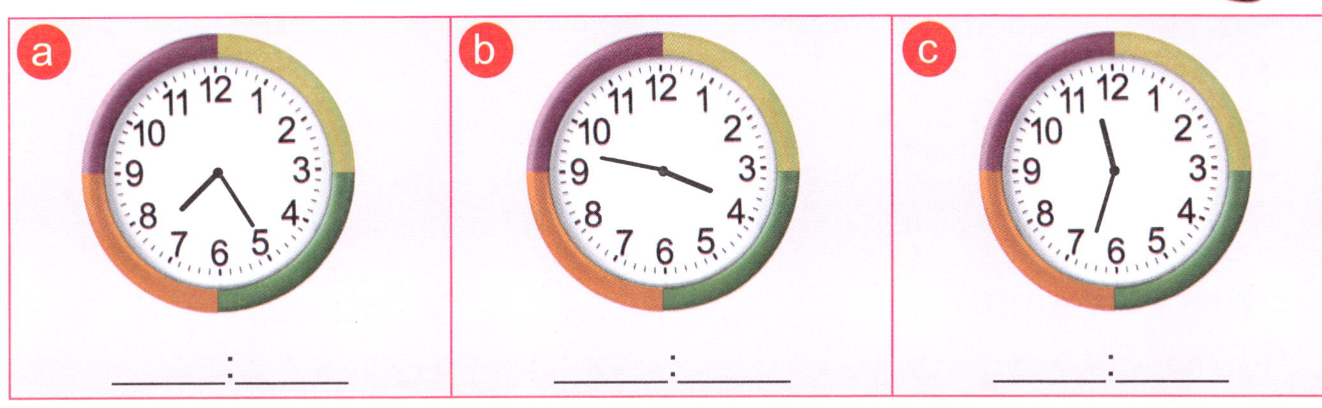

2. What is the time if hour hand is between 9 and 10, and minute hand is 1 small division beyond 5? ______________________________

3. A clock shows 5:19. Hour hand is between ________ and ________. Minute hand is at ________ divisions beyond ________.

4. Draw hands on the clock face to show the indicated time.

Worksheet - 36

Time in am and pm, 24-hour Clock Time in Seconds

1. **Write the time in am or pm.**

   a) Half past 8 in the morning _____ b) 15 minutes to 10 in the night _____

2. **What time will it be?**

   a) 1 hour after 12 midnight _____ b) 6 hours after 4:30 am _____

3. **What time was it?**

   a) 5 hours before 12 noon _____ b) 1 hour before 1:50 am _____

4. **Change to 24-hour clock times.**

   a) 4 am _______________ b) 8 pm _______________

   c) 12 noon _______________ d) 12 midnight _______________

5. **Change to 12-hour clock times.**

   a) 0300 hours _______________ b) 1630 hours _______________

6. **Convert into seconds.**

   a) 2 minutes 43 seconds

   ____

   b) 2 hours 25 minutes 30 seconds

   ____

7. a) A faulty clock gains 1 minute in every hour. If it was set right at 10:00 am what time will it show when it is actually 7:00 pm on the same day?

   ____

   b) A faulty clock loses 3 minutes in every hour. It was set right at 12 noon on Monday. What will be the day and time when it will be one hour late?

   ____

Worksheet - 37

Temperature

1. Fill in the blanks.

a) Temperature can be measured using a ____________________.

b) Two scales used for measuring temperature are ____________ scale and ____________________ scale.

c) Temperature of 50 on Celsius scale is written as ____________________.

d) Temperature of 50 on Fahrenheit scale is written as ____________________.

e) 100°F is ____________________ (hotter / colder) than 100°C.

f) 80°C is ____________________ (hotter / colder) than 80°F.

g) Normal temperature of a human body is ____________________ °F.

2. Weather of a place depends on the temperature range of that place.

Range of temperature	Type of weather
Below 0°C	Very cold
0°C – 10°C	Cold
10°C – 20°C	Cool
20°C – 25°C	Moderate
25°C – 30°C	Warm
30°C – 40°C	Hot
More than 40°C	Very hot

State the kind of weather a place will have if temperature is:

a) 14°C ____________ b) 26°C ____________

c) 2°C ____________ d) 45°C ____________

e) 21°C ____________ f) 36°C ____________

g) Weather of a place is very hot. Its temperature will be more than ____________ °C.

h) Weather of a place is cool. Its temperature will be between __________ °C and __________ °C.

Unit-8
Money

Worksheet - 38

Conversion of Dollars and Cents
Operations on Money

1. Convert into cents.

a) $ 35.15 = __________ cents   b) $ 26.06 = __________ cents

2. Convert into dollars.

a) 3175 cents = $ __________   b) 44971 cents = $ __________

3. Solve the following:

(a) Add $ 34.98 and $ 142.90	(b) Subtract $ 34.98 from $ 237.95
(c) Multiply $ 235 by 15	(d) Divide $ 4520 by 20

Worksheet - 39

Word Problems

1. Ria bought fruits for $ 436.50 and vegetables for $ 213.50. What amount did she spend?

2. I had $ 19,000 with me. I bought a camera for $ 3,500. What amount was left with me?

3. Cost of one luxury car is $ 33,500. What will be the cost of 16 such luxury cars?

4. Cost of 13 items is $ 910,000. What is the cost of each item?

Worksheet - 40

Preparing Bills

ABC FAST FOODS

Rate List

Packet of Chips	: $ 2	Packet of Biscuits	: $ 3	Sandwich	: $ 5
Burger	: $ 5	Pizza	: $ 12	Spring Roll	: $ 2
Cola	: $ 1	Juice	: $ 3	Coffee	: $ 5

Monika ordered the following items for her birthday party:

4 packets of chips, 2 packets of biscuits, 4 sandwiches, 3 burgers, 2 pizzas, 4 bottles of cola, and 1 can of juice.

Prepare a bill for her order.

ABC FAST FOODS

Bill

Date: ____________

Item	Quantity	Rate	$
______________	______	______	______
______________	______	______	______
______________	______	______	______
______________	______	______	______
______________	______	______	______
______________	______	______	______
______________	______	______	______
			Total $ = _____

She gave a note of $ 100. What amount did she get back? $ ___________

Point, Line Segment, Line and Ray

Fill in the blanks.

1. A line segment has ____________ end-points and a definite ____________.
2. If a line segment is stretched on both sides we get a ____________.
3. Three or more points are said to be ____________ if they lie on the same line.
4. We cannot find the length of ray AB as it extends ____________ in one direction.
5. An infinite number of ____________ can be drawn to pass through a fixed point.
6. Through three given points a maximum of ____________ line(s) can be drawn.
7. A, B, C and D are four points such that ABCD is a quadrilateral. How many line segments can be drawn using these points? ____________
8. A point determines ____________.
9. Line segment AB can also be named as line segment ____________.
10. The symbol ____________ represents line AB.
11. We cannot find the length of line as it extends ____________ on both sides.
12. If a line segment is stretched only on one side we get a ____________.
13. The starting point of the ray is also called its ____________ point.
14. Ray AB and ray BA are not the same as they point in ____________ directions.
15. Through ____________ given points only one line can be drawn.

Worksheet - 42

Angles

Fill in the blanks.

1. Two different rays with same initial point form an __________.
2. The two rays forming an angle are called the __________ of the angle.
3. The common initial point of the two rays (arms) is called the __________ of the angle.
4. a) Name of the angle: ∠ ______________
   b) Arms: ______________
   c) Vertex: ______________

C
A
B

5. a) How many angles are formed in the given figure?
   b) Name each angle. ______________________
   ____

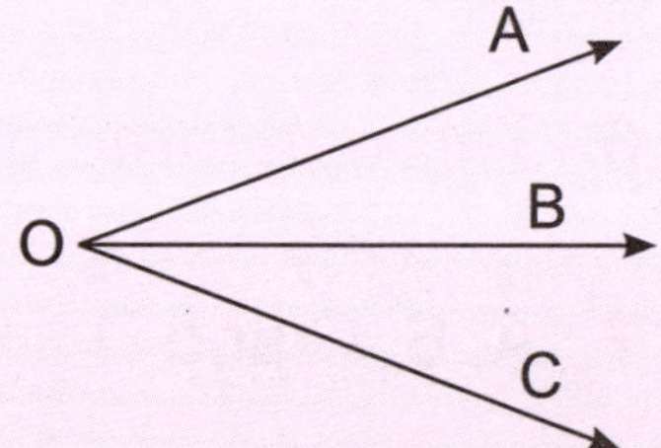

6. Measure each of the following angles with a protractor:

a

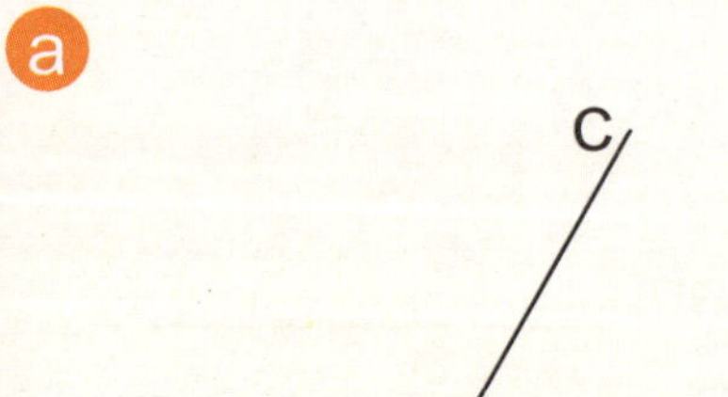

∠ ____ = ____°

b

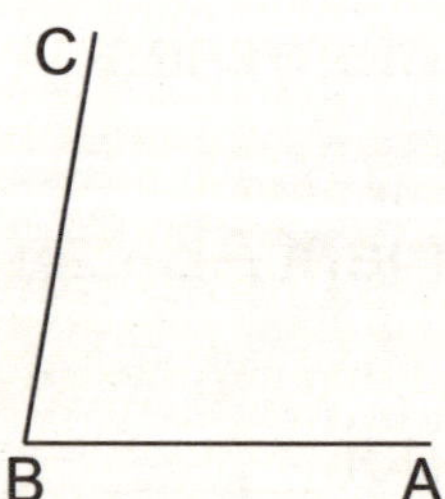

∠ ____ = ____°

c

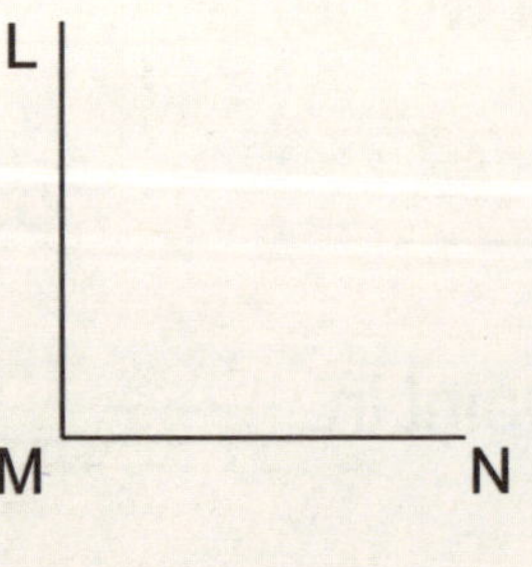

∠ ____ = ____°

d

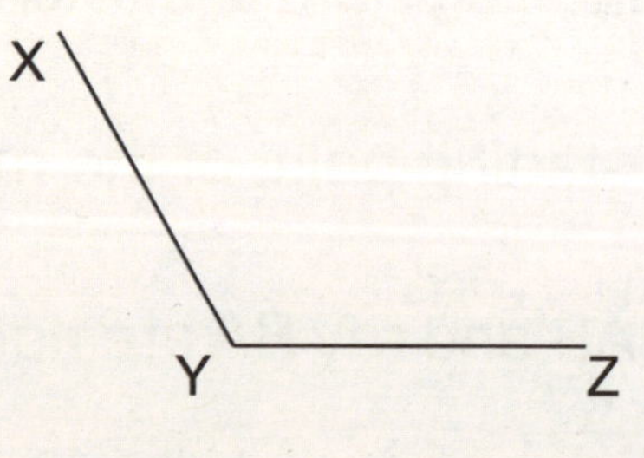

∠ ____ = ____°

Worksheet - 43

Constructing Angles Using a Protractor
Kinds of Angles

1. Construct the following angles using a protractor:

a 35°	b 90°
c 148°	d 210°

2. Fill in the blanks.

a) An angle whose measure is between 0° and 90° is called an ______ angle.

b) An angle whose measure is 90° is called a ______ angle.

c) An angle whose measure is between 90° and 180° is called an _____ angle.

d) An angle whose measure is 180° is called a ______ angle.

3. Classify the angles as acute, right or obtuse.

a) 88° ________________ angle   b) 90° ________________ angle

c) 95° ________________ angle   d) 180° ________________ angle

Worksheet - 44

Circles and Polygons
Rotation About Given Axis

1. In the given figure:

   a) OC is ______________.

   b) AB is ______________.

   c) DE is a ______________.

   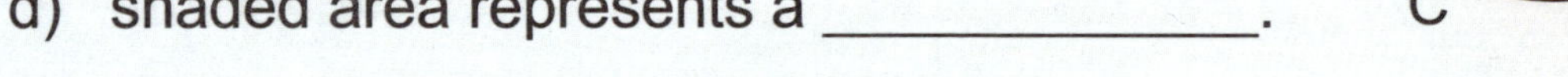

   d) shaded area represents a ______________.

2. If radius of a circle is 6 cm, its diameter is ______________ cm.

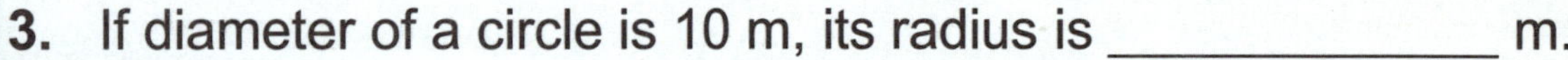

3. If diameter of a circle is 10 m, its radius is ______________ m.

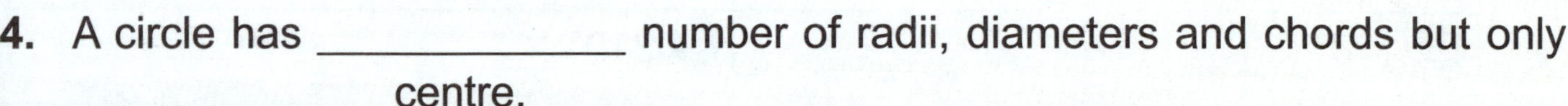

4. A circle has ______________ number of radii, diameters and chords but only ______________ centre.

5. Identify the following polygons.

a) 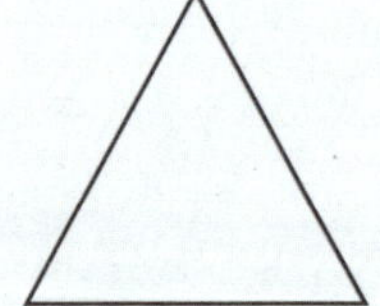 b) 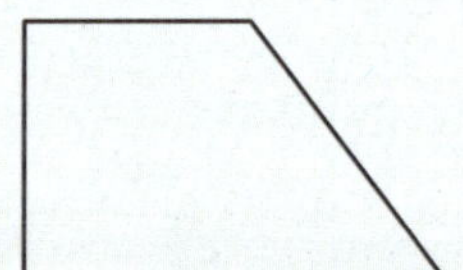 c) 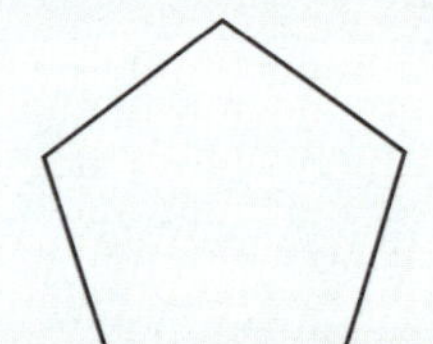 d) 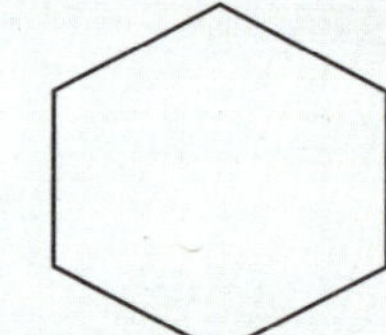

______________ ______________ ______________ ______________

6. Draw the figure obtained by rotating about the shown axis.

a 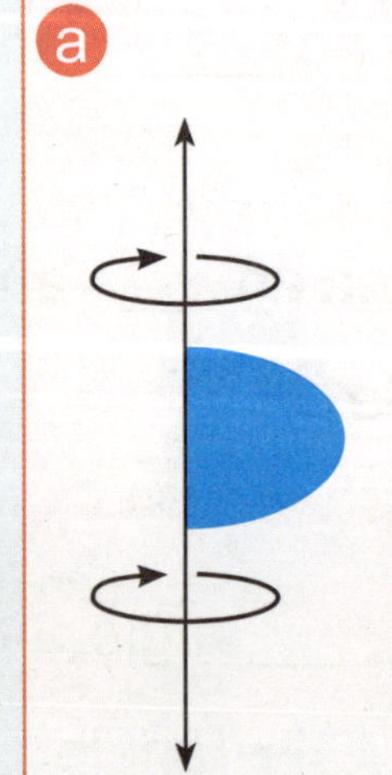

b 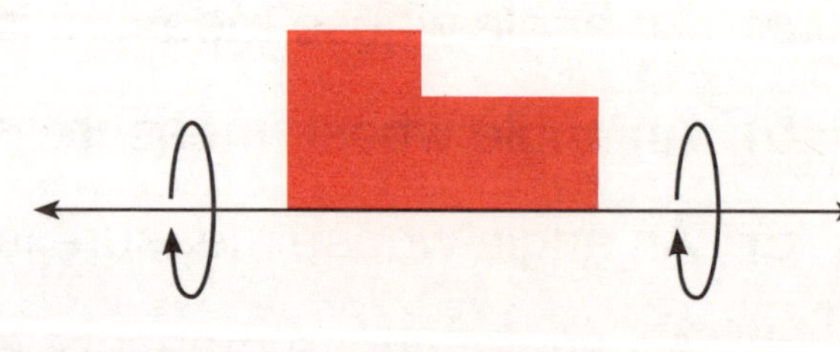

Worksheet - 45

Perimeter, Area of Square and Rectangle

1. Side of a square is 12 m. Find its perimeter and area.

____

____

____

2. Length of a rectangle is 11 cm and its breadth is 9 cm. Find its perimeter and area.

____

____

____

3. A square garden has a side of 30 m. A rectangular garden measures 15 m by 10 m. Which has greater perimeter and by how many metres?

____

____

____

4. A square plot has a side of 30 m. A rectangular plot measures 50 m by 12 m. Which has greater area and by how many square metres?

____

____

____

Worksheet - 46

Volume of a Cube and Cuboid

1. Side of a cube is 10 cm. Find its volume.

2. A solid cuboid is 20 cm × 10 cm × 5 cm. Find its volume.

3. Side of a cube is 8 cm. Dimensions of a cuboid are 12 cm, 8 cm and 6 cm. Which has greater volume and by how many cubic centimetres?

4. Volume of a cube is 1000 cubic cm. What is the length of its each side?

Worksheet - 47

Repeating, Increasing and Reducing Patterns

Draw the next three figures for each of the following:

a						
b						
c						
d						
e						
f						
g						
h						
i						

Worksheet - 48

Number, Letter and Code Patterns

Write the next three terms for each of the following:

a	11, 13, 17, 19,	
b	120, 240, 360, 480,	
c	117, 104, 91, 78,	
d	101, 21012, 3210123, 432101234,	
e	ABZ, CDY, EFX, GHW,	
f	PABC, QDEF, RGHI, SJKL,	
g	A, BC, DEF, GHIJ,	
h	ACE, EGI, IKM,	
i	AA11, BB22, CC33, DD44,	
j	B9A , D8C, F7E, H6G,	

Worksheet - 49

Drawing Pictographs and Bar Charts

1. The ages of all people working in an office are as follows: 38, 43, 25, 39, 36, 39, 44, 39, 44, 46, 54, 33, 56, 24, 53, 38, 38, 46, 43 and 22.

Arrange these observations in the table given below.

Age group	Tally bars	Number of employees
21 – 30 years		
31 – 40 years		
41 – 50 years		
51 – 60 years		

2. **Now represent this data by a bar chart.**

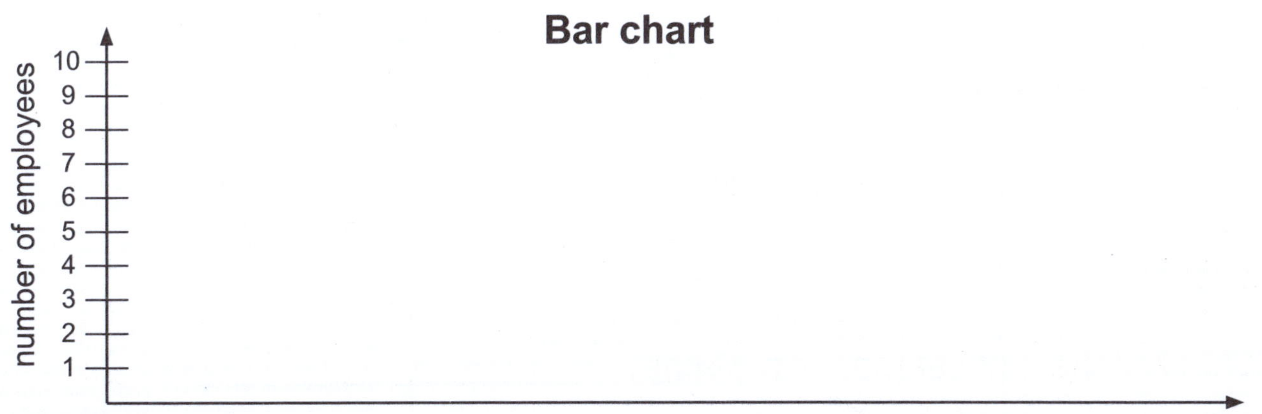

3. **Draw a pictograph. [Take ☺ = 1 person]**

Age group	Number of employees
21 – 30 years	
31 – 40 years	
41 – 50 years	
51 – 60 years	

Worksheet - 50

Reading Tables

Ann, Sam, Sonia, Paul and Maggie are classmates. They decide to compare their results by writing them on the board.

	English (50)	Science (50)	Maths (50)	History (50)	Total (200)
Ann	40	49	36	40	________
Sam	39	35	48	37	________
Sonia	40	42	43	39	________
Paul	42	43	44	38	________
Maggie	43	39	40	42	________

Using above information fill in the blanks.

1. What is the highest total? ____________________

Who has this total? ____________________

2. Which two students have the same total? ____________ and ____________

What is this total? ____________________

3. What are the highest marks in Maths? ____________________

Who scored these marks? ____________________

4. What did Paul score in English? ____________________

5. What was Sonia's score in History? ____________________

Was she the highest scorer? (yes / no) ____________________

6. What were the highest marks in Science? ____________________

Who scored these marks? ____________________

Unit-1 Big Numbers

Worksheet-1

1. 2 millions 3 hundred thousands 4 ten thousands 5 thousands 9 hundreds 0 tens and 7 ones
2. 9 ten millions 8 millions 7 hundred thousands 3 ten thousands 4 thousands 2 hundreds 1 ten and 5 ones
3. a) 3 760 591 b) 26 427 036
4.

	TM	M	H-Th	T-Th	Th	H	T	O
a)		8	3	7	3	9	5	8
b)	8	9	3	4	7	9	2	7

5. a) 3 241 368 b) 82 022 388

Worksheet-2

1. a) Five million two hundred thirty-nine thousand eight hundred one
   b) Eighty-seven million two hundred ninety-seven thousand two hundred fifty-eight
2. a) 7,738,401 b) 30,120,280
3. a) 7000000 + 200000 + 10000 + 4000 + 700 + 50
   b) 20000000 + 9000000 + 900000 + 9000 + 900 + 9
4. a) 2580276 b) 90947019
5. 1287452
6. 12764400, 12764401
7. 75212092
8. 7012399, 7012398
9. 5647249, 5647250
10. a) 3 200 000, 5 200 000 b) 33 000 000, 43 000 000

Worksheet-3

1. 6446316
2. 19443741
3. 5477997, 5484572, 5546499
4. 24285727, 24277420, 24273627
5. 7000000
6. 30000000
7. a) 7654321; 1234567 b) 88888888; 10000000
8. 9864210
9. 25252525
10. 3577999
11. 28752000
12. 4157398
13. 1000000
14. 5555555

Worksheet-4

1. a) zero b) I, V, X, L, C, D and M
   c) V, L, D and M d) I and X e) 10,000
2. a) XLVIII b) XCVII c) CLXXXII
   d) CCCLXIII c) DCXXIV f) DCCXXXV
   g) DCCCXIX h) MCCCLVI
3. DDXV (✗) CMLIX (✓) LVXIII (✗)
   CDLXVI (✓) VXII (✗) MMDXL (✓)
4. a) 25 b) 46 c) 139 d) 218
   e) 661 f) 1663 g) 1693 h) 2714
5. a) LXV b) DXL

Unit-2 Operations on Numbers

Worksheet-5

1. a) 5956987 b) 86646759 c) 6092582 d) 67303645
2. a) 5341111 b) 51123403 c) 1598789 d) 3724590

Worksheet-6

1. 10,412,348
2. 2,700,728; 33,044
3. 79,070
4. $ 12,107,400

Worksheet-7

1. 7128790
2. correct
3. incorrect
4. a) 1223272 b) 1282372 c) 47847593; 54742632
   d) 517004 e) 178463 f) 2531287
   g) 45281352 h) 53618679 i) 0
   j) 458352 k) 32351023 l) 1921376

Worksheet-8

a) 2684208 b) 30696993 c) 8145102
d) 24011039 e) 2950531 f) 86666775
g) 8567406 h) 13998798

Worksheet-9

1. 702512
2. 362416
3. 265654; 4
4. 680439; 4

Worksheet-10

1. 656,100
2. 259,700
3. 84,683
4. 84,306; 20

Worksheet-11

1. a) 24132000 b) 6530000 c) 30000000
   d) 50000000 e) 8844000 f) 10550000
   g) 1320000 h) 1350000 i) 11200000
   j) 10560000 k) 6,000,000 l) 12,000,000
2. a) 9642458 b) 7468521 c) 3485647
   d) 874141 e) 123908; 456120
   f) 621248 g) 5325; 58743
   h) 1234; 123400; 1234000 i) 0
   j) 0 k) 2906000 l) 0

Worksheet-12

1. a) 8000000 b) 2000000
2. a) 40000000 b) 10000000
3. 9800000
4. 14000000
5. 8
6. 158
7. 110
8. $ 1,000,000
9. 60,000,000

Worksheet-13

1. 8
2. 15
3. 17
4. 12
5. 3

Unit-3 Multiples and Factors

Worksheet-14

1. 14, 28, 42, 56, 70 2. 36, 45, 54, 63
3. a) No b) Yes
4. 19, 38, 59(✗), 76, 95, 113(✗), 134(✗), 152, 171, 190
5. a) 0 or 2 or 4 or 6 or 8 b) 2
   c) 1 or 3 or 5 or 7 or 9 d) 2
6. 9013 (✓), 1265 (✓), 9826 (✗), 12188 (✗), 23110 (✗)
7. 10, 12, 14, 16, 18 8. 101, 103, 105, 107, 109
9. 3097, 3099, 3101, 3103, 3105
10. 22, 44, 66 11. a) True b) False

Worksheet-15

1. a) factor b) 96 c) a, b, c; d d) 10, 8, 3
   e) factor f) 1, 3, 9 g) 1, 3, 5, 9, 15, 45
   h) 1, 5 i) factor j) 7 k) 13, 8; 104
   l) 7; 28
2. a) True b) False c) True d) False
   e) True f) False g) True h) True

Worksheet-16

1. a) prime b) 15 c) 25 d) even; prime
   e) 2, 3 f) 37 – 7 g) 43 h) composite
   i) 91, 92, 93, 94, 95, 96, 98, 99
2. a) 1962; 9464 b) 3153; 9327 c) 7544; 3116; 7636
   d) 4435; 5380; 51445 e) 4392; 3192; 33432
   f) 4230; 5292; 7065; 4140 g) 7410; 4520; 5460
3. 0, 2, 4, 6, 8 4. 2, 5, 8 5. 0, 4, 8
6. 0, 5 7. 0, 1, 2, 3, 4, 5, 6, 7, 8, 9

Worksheet-17

1. $2 \times 2 \times 2$ 2. $2 \times 2 \times 7$ 3. $2 \times 2 \times 13$
4. $2 \times 2 \times 2 \times 3 \times 3$ 5. $2 \times 2 \times 2 \times 2 \times 5$ 6. $2 \times 2 \times 5 \times 5$

Worksheet-18

1. a) 12 b) 30 2. 20
3. a) 90 b) 120
4. 5 boxes of brand 'A', 8 boxes of brand 'B' 5. 12:00 noon

Unit-4 Fractions

Worksheet-19

1. $\frac{9}{19}$(✗), $\frac{11}{2}$(✓), $\frac{25}{7}$(✓), $\frac{3}{9}$(✗), $\frac{8}{17}$(✗), $\frac{105}{46}$(✓), $\frac{199}{220}$(✗)
2. a) $2\frac{1}{4}$ b) $3\frac{8}{9}$ c) $55\frac{1}{56}$
3. a) $\frac{23}{6}$ b) $\frac{13}{5}$ c) $\frac{35}{8}$ d) $\frac{19}{7}$ e) $\frac{78}{7}$ f) $\frac{28}{9}$
4. a) $2\frac{5}{6}$ b) $1\frac{3}{8}$ c) $1\frac{4}{9}$ d) $3\frac{1}{7}$ e) $11\frac{1}{9}$ f) $15\frac{5}{13}$
5. a) like b) like c) unlike d) like e) unlike f) unlike
6. a) $\frac{9}{5}$ b) $\frac{6}{11}$ c) $\frac{3}{5}$ d) $\frac{14}{1}$ e) $\frac{101}{111}$ f) $\frac{29}{89}$

Worksheet-20

1. a) 3/4 b) 8/10 or 4/5 c) 9/10 d) 76/70 or 38/35
2. a) 2/5 b) 2/10 or 1/5 c) 7/24 d) 5/24
3. a) 1/14 b) 9/22 c) $1\frac{33}{289}$ d) 0
4. a) 5/4 b) 27/8 c) 2/7 d) 7/51
5. a) 25/2 b) 3/16 c) 45/91 d) 4/39

Worksheet-21

1. 5/8 2. 4/45 m 3. 6/20 or 3/10 4. 64

Unit-5 Decimals

Worksheet-22

1. a) zero point five b) ten point zero six
   c) two point zero four three
2. a) 0.2 b) 1.3 c) 0.21 d) 1.09
   e) 0.007 f) 3.123
3. a) 2/10 b) $1\frac{43}{100}$ c) $1\frac{463}{1000}$ d) $2\frac{14}{1000}$
4.

	Hundreds 100	Tens 10	Ones 1	Tenth $\frac{1}{10}$	Hundredths $\frac{1}{100}$	Thousands $\frac{1}{1000}$
a)			6	2		
b)		1	8	5	7	
c)	1	2	3	0	5	2

5. a) 1.1 b) 1.46 c) 7.176
6. a) 4 + 6 /10 b) 1 + 2/100

Worksheet-23

1. a) like b) unlike
2. a) True b) False c) False d) False
3. a) 4.30, 0.84 b) 23.620, 2.937
   c) 1.900, 0.002 d) 5.87, 7.90
4. a) 15.6 b) 8.853
5. a) 7.8 b) 8.809

Worksheet-24

1. a) 31.44 b) 2.091 2. a) 9.3 b) 6.43
3. a) 75.2 b) 491.6 c) 3934
   d) 0.62 e) 2.248 f) 0.0125

Worksheet-25

1. 63 2. 295 3. 308 4. 2315 5. 12934
6. 2550 7. 108 8. 3995 9. 4650 10. 296
11. 3052 12. 12800 13. 8.19 14. 3.490 15. 41.65

Unit-6 Metric Measures

Worksheet-26

1. metre; m 2. centimetre; cm 3. 100
4. millimetre; mm 5. 10 6. metre; m
7. 1000 8. kilogram; kg 9. gram; g 10. 1000
11. litre; L 12. millilitre; ml 13. 1000 14. divide
15. multiply 16. 10 17. quintal 18. 1000 19. 1000

Worksheet-27

1. 4 2. 6 3. 12 4. 46 5. 590
6. 80 7. 5 8. 4 9. 5 10. 9
11. 60 12. 24000 13. 300 14. 6000 15. 4
16. 2000 17. 5 18. 2000 19. 2 20. 18
21. 6000 22. 23

Worksheet-28

1. 75 cm 2. 7.37 m 3. 14.038 km 4. 795 ml
5. 6.84 L 6. 25.999 kl 7. 869 mg 8. 31.655 g
9. 27.962 kg

Worksheet-29

1. 24 cm 2. 1.23 m 3. 56.02 km
4. 565 ml 5. 1.883 L 6. 0.129 kl
7. 292 mg 8. 7.1 g 9. 4.105 kg

Worksheet-30

1. 375 cm 2. 16.64 m 3. 236 km
4. 3025 ml 5. 491.96 L 6. 51975 kl
7. 2268 mg 8. 15600 g 9. 178125 kg

Worksheet-31

1. 21 cm 2. 2.925 m 3. 23.5 km
4. 83 ml 5. 6.25 L 6. 12.8 kl
7. 11 mg 8. 117.5 g 9. 329 kg

Worksheet-32

1. 931 km 2. 41.5 km 3. 8 m 4. 75 cm

Worksheet-33

1. 12 tons 2. 8 q 3. 250 g 4. 500 mg

Worksheet-34

1. 11 kl 2. 4 kl 3. 75 L 4. 2 L

Unit-7 Time and Temperature

Worksheet-35

1. a) 7:25 b) 3:47 c) 11:33
2. 9:26 3. 5; 6; 4; 3
4. a) b)  c) 
d)  e)  f) 

Worksheet-36

1. a) 8:30 am b) 9:45 pm
2. a) 1:00 am b) 10:30 am
3. a) 7:00 am b) 12:50 am
4. a) 0400 hrs b) 2000 hrs c) 1200 hrs d) 0000 hrs
5. a) 3:00 am b) 4:30 pm
6. a) 163 sec b) 8730 sec
7. a) 7:09 pm b) 8:00 am; Tuesday

Worksheet-37

1. a) thermometer b) Celsius; Fahrenheit
   c) 50°C d) 50°F e) colder
   f) hotter g) 98.6
2. a) Cool b) Warm c) Cold d) Very hot
   e) Moderate f) Hot g) 40 h) 10; 20

Unit-8 Money

Worksheet-38

1. a) 3515 b) 2606
2. a) 31.75 b) 449.71
3. a) $ 177.88 b) $ 202.97 c) $ 3525 d) $ 226

Worksheet-39

1. $ 650 2. $ 15,500 3. $ 536,000 4. $ 70,000

Worksheet-40

ABC FAST FOODS

Bill

Date: ____________

Item	Quantity	Rate	$
Chips	4	2	8
Biscuit	2	3	6
Sandwich	4	5	20
Burger	3	5	15
Pizza	2	12	24
Cola	4	1	4
Juice	1	3	3
			Total $ = 80

She gets back $ 20.

Unit-9 Geometry

Worksheet-41

1. two; length 2. line 3. collinear
4. infinitely 5. lines 6. three
7. six 8. position 9. BA
10. $\overleftrightarrow{AB}$ 11. infinitely 12. ray
13. initial 14. opposite 15. two

Worksheet-42

1. angle 2. arms 3. vertex
4. a) ∠ABC; BA and BC; B
5. a) three b) ∠AOB, ∠BOC, ∠AOC
6. a) ∠ABC = 60° b) ∠ABC = 80°
   c) ∠LMN = 90° d) ∠XYZ = 120°

Worksheet-43

1. a) b) c) d) 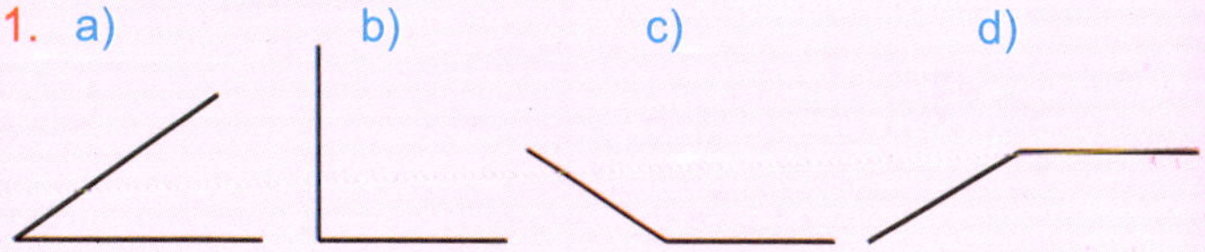
2. a) acute b) right c) obtuse d) straight
3. a) acute b) right c) obtuse d) straight

Worksheet-44

1. a) radius b) diameter c) chord d) sector
2. 12
3. 5
4. infinite; one
5. a) triangle b) quadrilateral c) pentagon d) hexagon
6. a) b)

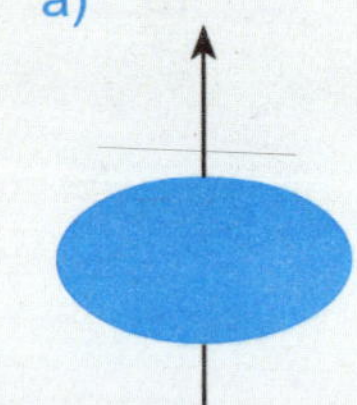

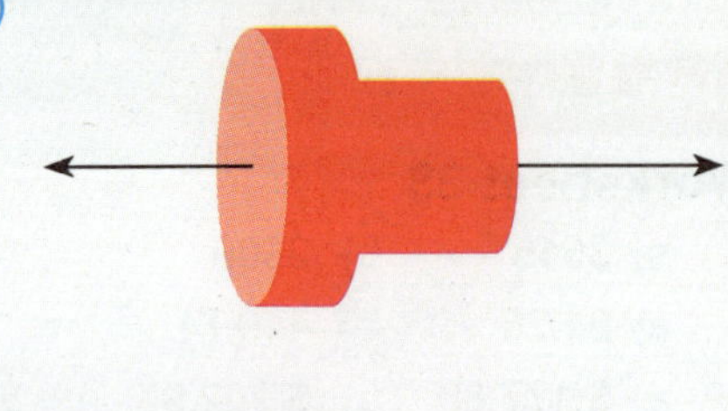

Unit-10 Perimeter, Area and Volume

Worksheet-45

1. 48 m; 144 sq. m
2. 40 cm; 99 sq. cm
3. square garden; 70 m
4. square plot; 300 sq. m

Worksheet-46

1. 1000 cu. cm
2. 1000 cu. cm
3. cuboid; 64 cu. cm
4. 10 cm

Unit-11 Patterns

Worksheet-47

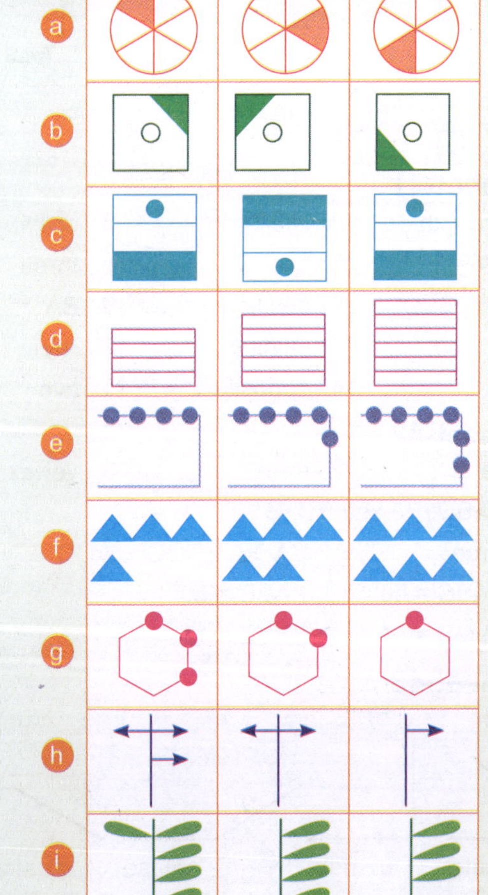

Worksheet-48

a) 23, 25, 29
b) 600, 720, 840
c) 65, 52, 39
d) 54321012345, 6543210123456, 765432101234567
e) IJV, KLU, MNT
f) TMNO, UPQR, VSTU
g) KLMNO, PQRSTU, VWXYZAB
h) MOQ, QSU, UWY
i) EE55, FF66, GG77
j) J5I, L4K, N3M

Unit-12 Data Handling

Worksheet-49

1.

Age group	Tally bars	Number of employees
21 – 30 years	|||	3
31 – 40 years |卌 |||	8
41 – 50 years	卌		6
51 – 60 years	|||	3

2.

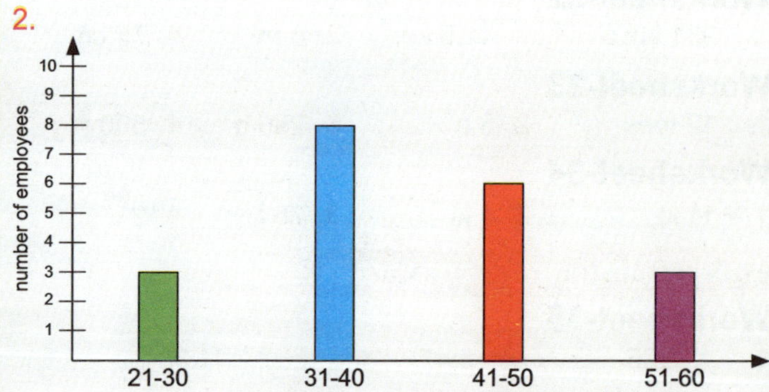

3.

Age group	Number of employees
21 – 30 years	☺ ☺ ☺
31 – 40 years	☺ ☺ ☺ ☺ ☺ ☺ ☺ ☺
41 – 50 years	☺ ☺ ☺ ☺ ☺ ☺
51 – 60 years	☺ ☺ ☺

Worksheet-50

	Ann	Sam	Sonia	Paul	Maggie
Total (200)	165	159	164	167	164

1. 167; Paul
2. Sonia and Maggie; 164
3. 48; Sam
4. 42
5. 39; no
6. 49; Ann

Printed in India